Short Cases in
Clinical Biochemistry

Short Cases in Clinical Biochemistry

Geoffrey V. Gill
MSc, MD, MRCP (UK), DTM & H
Senior Registrar,
Department of Clinical Biochemistry and
Metabolic Medicine,
University of Newcastle upon Tyne,
Royal Victoria Infirmary,
Newcastle upon Tyne, UK

Michael F. Laker
MD, MRCPath, Dip Biochem
Senior Lecturer,
Department of Clinical Biochemistry and
Metabolic Medicine,
University of Newcastle upon Tyne,
Royal Victoria Infirmary,
Newcastle upon Tyne, UK

K. George M. M. Alberti
MB, DPhil, MRCPath, FRCP
Professor,
Department of Clinical Biochemistry and
Metabolic Medicine,
University of Newcastle upon Tyne,
Royal Victoria Infirmary,
Newcastle upon Tyne, UK

CHURCHILL LIVINGSTONE
EDINBURGH LONDON MELBOURNE AND NEW YORK 1984

CHURCHILL LIVINGSTONE
Medical Division of Longman Group Limited

Distributed in the United States of America by
Churchill Livingstone Inc., 1560 Broadway, New
York, N.Y. 10036, and by associated companies,
branches and representatives throughout the
world.

© Longman Group Limited 1984

All rights reserved. No part of this publication may
be reproduced, stored in a retrieval system, or
transmitted in any form or by any means,
electronic, mechanical, photocopying, recording or
otherwise, without the prior permission of the
publishers (Churchill Livingstone, Robert
Stevenson House, 1–3 Baxter's Place, Leith Walk,
Edinburgh EH1 3AF).

First published 1984

ISBN 0 443 02929 6

British Library Cataloguing in Publication Data
Gill, Geoffrey V.
   Short cases in clinical biochemistry.
   1. Biological chemistry   2. Chemistry, Clinical
   I. Title   II. Laker, Michael F.
   III. Alberti, K.G.M.M.
   612'.015   QP514.2

Library of Congress Cataloging in Publication Data
Gill, Geoffrey V.
   Short cases in clinical biochemistry.
   1. Chemistry, Clinical — Case studies.
   2. Chemistry, Clinical — Examinations,
   questions, etc. I. Laker, Michael F. II. Alberti,
   K. G. M. M. (Kurt George Matthew Mayer)
   III. Title.
   RB40.G56   1984   616.07'56'0926
   83-14987

Printed in Singapore by Selector Printing Co Pte Ltd

Preface

To many doctors, clinical biochemistry laboratories are jungles of modern technology, where automation and quality control seem to take precedence over all else. To the clinical biochemist, however, clinicians are often distant voices on the phone demanding ever increasing tests, sometimes of doubtful value. The reason is poor undergraduate and postgraduate education, and a lack of communication on both sides. The situation is particularly unfortunate as biochemical reports require active interpretation on the wards (unlike marrows, cultures and biopsies), and this is a skill in which clinicians often have little expertise.

Fortunately, the times are changing. Younger clinicians are learning to use laboratories sensibly, and clinical biochemists are now taking their specialty back to the wards and the bedside.

In an attempt to help this closer liasion, we have collected together the laboratory reports which are presented in this short book. They are all taken from our own files and cover many common problems seen in patients and laboratories. Each is presented as a question and answer, with short explanatory notes. We hope that this approach will be of use to interested medical students, doctors, biochemists and medical laboratory scientific officers. In particular it should be suitable for those unfortunate souls preparing for MRCP, MRCPath and MCB examinations.

Finally, we would like to thank many people—both in the laboratory and on the wards—for their help and stimulation to this project. In particular we are grateful to Gillian Atkins for secretarial assistance and Peter Baylis for endocrine advice.

Newcastle upon Tyne, 1984
G. V. G.
M. F. L.
K. G. M. M. A.

Reference Ranges (relevant to cases in book)

Electrolytes and tests of renal function (plasma)

Sodium	134–147	mmol/l
Potassium	3.5–5.0	mmol/l
Chloride	96–106	mmol/l
Carbon dioxide	22–29	mmol/l
Urea	2.5–7.0	mmol/l
Creatinine	53–124	µmol/l
Creatinine clearance	males 117–170	ml/min
	females 104–158	ml/min
Urinary protein	< 200	mg/24 hours

figures for young adults, varies with age, sex, surface area and pregnancy.

Blood gases and derived indices (arterial blood)

pH	7.35– 7.42	
pO_2	12.0 –14.0	kPa
pCO_2	4.5 – 6.1	kPa
Base excess	–3.0 to + 3.0	mmol/l
Standard bicarbonate	21.0 –25.0	mmol/l
Actual bicarbonate	21.0 –25.0	mmol/l

Proteins and tests of liver function (serum)

Total protein	60–80	g/l
Albumin	34–50	g/l
Bilirubin	2–17	µmol/l
Alkaline phosphatase	20–90	U/l
Aspartate aminotransferase	4–20	U/l
Gamma glutamyl transpeptidase	males < 60	U/l
	females < 35	U/l

Tests of calcium metabolism (serum)

Calcium	2.12–2.62	mmol/l
Phosphate	0.8 –1.4	mmol/l
Magnesium	0.7 –1.0	mmol/l

Lipids (fasting serum)

Cholesterol	3.6 –7.3	mmol/l
Triglycerides	0.45–1.80	mmol/l

Enzymes (serum)

Creatine kinase	0–50	U/l
Amylase	70–300	U/l
Lactate dehydrogenase	50–220	U/l

Miscellaneous

Haemoglobin	males	13.5–18.0	g/l	blood
	females	11.5–16.5	g/l	blood
White cell count		4.0–7.0 × 10^9	mm^{-3}	blood
Haemoglobin A_1		5.0–7.5%		
Blood glucose (fasting)		3.5–5.0	mmol/l	blood
Vitamin B_{12}		160–1000	ng/l	serum
Folate		>4	µg/l	serum
Osmolality		282–298	mmol/kg	plasma
HMMA		<40	µmol/24 hrs	urine
HIAA		<60	µmol/24 hrs	urine
Urate	males	0.15–0.42	mmol/l	serum
	females	0.12–0.39	mmol/l	serum

Porphyrin metabolism (urine and faeces)

Aminolaevulinic acid	0–40	µmol/24 hrs	urine
Porphobilinogen	0–16	µmol/24 hrs	urine
Uroporphyrin	0–49	µmol/24 hrs	urine
Coproporphyrin (urine)	0–430	nmol/24 hrs	urine
Protoporphyrin	0–200	nmol/g dry weight	faeces
Coproporphyrin (faeces)	0–76	nmol/g dry weight	faeces

Hormones (serum)

Thyroxine		60–150	nmol/l
Triiodothyronine		1.2–3.0	nmol/l
Free thyroxine index		50–160	
TSH		<6.0	mU/l
Cortisol	9 am	190–720	nmol/l
	12 MN	<220	nmol/l
Urinary 11 OHCS		145–700	nmol/24 hours
Testosterone	males	8.5–27.5	nmol/l
	females	0.9–3.2	nmol/l
17 β oestradiol	males	80–230	pmol/l
	females	50–450 (pre-ovulation)	pmol/l
		450–1400 (ovulation peak)	pmol/l
		100–1000 (post-ovulation)	pmol/l
Progesterone	females	0.5–4.0 (pre-ovulation)	pmol/l
		20–80 (post-ovulation)	nmol/l
Insulin (fasting)		4–10	mU/l
FSH	males	1–7	U/l
	females	1–8 (follicular phase)	U/l
		0.5–3 (luteal phase)	U/l
		up to 15 (mid-cycle)	U/l
		>15 (post menopausal)	U/l
LH	males	1–6	U/l
	females	1–6 (follicular phase)	U/l
		1–3 (luteal phase)	U/l
		up to 25 (mid cycle)	U/l
		>20 (post menopausal)	U/l

GH (random)	doubtful use, levels > 20 exclude GH deficiency	mU/l
Prolactin	up to 450	mU/l
PTH	< 0.3 to 1.4	U/l

Abbreviations

ACTH	Adrenocorticotrophic hormone	HIAA	5-Hydroxy indoleacetic acid
ADH	Antidiuretic hormone	HMMA	4 Hydroxy 3-Methoxy Mandelic acid
ALA	Aminolaevulinic acid		
AP	Alkaline phosphatase	K	Potassium
AST	Aspartate transaminase	Ig	Immunoglobulin
BG	Blood glucose	LDH	Lactate dehydrogenase
Ca	Calcium	LFT	Liver function test
CK	Creatine kinase	LH	Luteinising hormone
Cl	Chloride	LRH	LH releasing hormone
CO_2	Carbon dioxide	Mg	Magnesium
pCO_2	Partial pressure of carbon dioxide	MI	Myocardial infarct
		Na	Sodium
CSF	Cerebrospinal fluid	pO_2	Partial pressure of oxygen
ECF	Extracellular fluid	11OHCS	11 hydroxy corticosteroids
ECG	Electrocardiograph	P	Phosphorus
ESR	Erythrocyte sedimentation rate	PBG	Porphobilinogen
		PTH	Parathormone
FSH	Follicle stimulating hormone	SIADH	Syndrome of inappropriate antidiuretic hormone
γGT	Gamma glutamyl transpeptidase	T_3	Triiodothyronine
		T_4	Thyroxine
GH	Growth hormone	TRH	Thyrotrophin releasing hormone
GTT	Glucose tolerance test		
HbA	Haemoglobin A	TSH	Thyroid stimulating hormone
HbC	Haemoglobin C		
HCO_3	Bicarbonate	U&E	Urea and electrolytes

1Q

An astute orthopaedic surgeon was seeing a middle-aged man newly referred with backache. He noted pallor of the mucous membranes and requested the following biochemical investigations, which returned as follows:

Total protein	94 g/l
Albumin	34 g/l
Serum calcium	3.51 mmol/l
Serum phosphate	1.10 mmol/l

What is the diagnosis?

2Q

There was some panic on a surgical ward when the following results were phoned back from the laboratory. They were routinely taken from a patient who had had a cholecystectomy the day before, and appeared to be doing well.

Na	K	Cl	CO_2	Urea	Creatinine
117 mmol/l	9.6 mmol/l	90 mmol/l	14 mmol/l	7.7 mmol/l	92 μmol/l

(Specimen not haemolysed)

What is the most likely cause?

1A

Myeloma

Backache and anaemia are both common presentations of this disease, and the high serum levels of globulin and calcium make the diagnosis almost certain. A classical 'M' band was seen on electrophoresis.

2A

'Drip arm' venepuncture

The patient had an infusion of 5% glucose with potassium chloride running into a drip entering the left hand. The harrassed house surgeon had taken the blood from the left antecubital fossa. A repeat specimen from the right arm gave normal results.

3Q

A 57-year-old man was being investigated for a peripheral neuropathy. He was not diabetic, had normal renal function, and he was not obviously malnourished. Serum levels of vitamin B_{12} and folate were:

    Folate   8 µg/l
    Vitamin B_{12} 1475 ng/l

The B_{12} result led to a clinical and biochemical line of enquiry which revealed the diagnosis.

What did the B_{12} level mean, and what was the cause of the man's neuropathy?

4Q

A 16-year-old girl developed progressive weakness of the arms and legs, and on admission was thought to have Guillan-Barré Syndrome. Because of increasing respiratory distress she was transferred to the intensive care unit, where blood gases were performed.

    pH   7.61
    pCO_2 1.8 kPa
    Base excess   + 4 mmol/l
    Standard bicarbonate 21 mmol/l
    Actual bicarbonate   13 mmol/l
    pO_2 15 kPa

What is the acid/base disturbance, and what is the likely cause?

3A

Alcoholic neuropathy

Grossly elevated serum B_{12} levels may be due to B_{12} therapy, leukaemia or hepatocellular dysfunction (the vitamin is present in very high concentrations in the liver). Subsequent liver function tests in this patient were abnormal, and it became apparent that the man was an alcoholic.

4A

Hysterical hyperventilation

The results indicate a respiratory alkalosis—entirely different from what would be expected with respiratory muscle weakness. The girl also had positive Trousseau's and Chvostek's signs (due to increased neuromuscular excitability secondary to a reduction in ionised calcium caused by the alkalosis). With sedation and firm handling the hyperventilation ceased, and she eventually made an uneventful recovery from her polyneuritis.

5Q

A 3-month-old infant with 'failure to thrive' had become non-specifically ill in the last few days. On admission the following routine area and electrolytes were returned:

Na	K	Cl	CO_2	Urea	Creatinine
125 mmol/l	5.9 mmol/l	92 mmol/l	17 mmol/l	13.5 mmol/l	56 μmol/l

What is the likely diagnosis?

6Q

A young man of 28 years was admitted for a meniscectomy, and mentioned to the houseman that he had had hepatitis as a child. The doctor ordered 'check LFTs' which came back abnormal, and remained so on repeat sampling. The operation was delayed pending a medical opinion.

Bilirubin	31 μmol/l
AST	15 U/l
Alkaline phosphatase	52 U/l

What is the likely diagnosis and how would you confirm it?

5A

Congenital adrenal hyperplasia

The results are typical of the 'salt losing' type, and the changes are identical with an Addisonian crisis. There is hyponatraemia, hyperkalaemia, metabolic acidosis and slight uraemia.

6A

Gilbert's syndrome

The physician who was consulted noted that the patient was asymptomatic and had no abnormal physical signs. He suspected Gilbert's syndrome. A differential bilirubin estimation showed that nearly all the circulating bilirubin was unconjugated, and after a 48 hour fast the total bilirubin had increased to 68 μmol/l—confirming the diagnosis. Later, the physician had a quiet word with the houseman about performing unnecessary tests! Gilbert's syndrome is due to an inherited disorder of bilirubin uptake into liver cells, and is typified by mild unconjugated hyperbilirubinaemia which is asymptomatic and increases on fasting.

7Q

A 60-year-old man was found to have myelomatosis. The following electrolyte abnormality was also noted:

Na	K	Cl	CO_2	Urea	Creatinine
125 mmol/l	3.3 mmol/l	94 mmol/l	22 mmol/l	2.5 mmol/l	50 μmol/l

What is the likely cause of this abnormality and how would you confirm it?

8Q

A middle-aged woman with diarrhoea thought that her symptoms were provoked by milk. She had a lactose tolerance test performed with the following results:

0 min	BG	4.0 mmol/l
50 g lactose given		
30 min	BG	4.1 mmol/l
60 min	BG	4.1 mmol/l
90 min	BG	4.4 mmol/l
120 min	BG	4.8 mmol/l

Is this abnormal, and if so what further steps would you take to establish a diagnosis?

7A

Pseudohyponatraemia

The high levels of paraprotein in myelomatosis may expand the plasma volume such that Na concentration falls, though the absolute Na content is the same. This effect may also be seen in severe hyperlipidaemia, and is distinguished from 'true' hyponatraemia by a normal plasma osmolality (this patient's plasma globulin level was 148 g/l and the plasma osmolality was 290 mmol/kg).

8A

Acquired lactase deficiency

A rise of less than 1 mmol/l of blood glucose is suggestive of lactose malabsorption. To exclude generalised malabsorption, a further tolerance test should be performed with 25 g glucose with 25 g galactose (the products of lactase action on lactose)—this should give a normal absorption curve. Enzyme analysis of small intestinal biopsy material, or breath hydrogen analysis after an oral lactose load, may also be used to make the diagnosis. These further investigations confirmed lactase deficiency to be present in this woman. No underlying cause could be found, but she responded well to a lactose-free diet.

9Q

A young couple were being investigated for primary subfertility. Both were clinically healthy, and the husband's semen was normal on 3 occasions. Temperature charts to demonstrate ovulation had been inconclusive. A number of blood tests were taken from the wife at one visit, and the form was marked 'one week prior to menstruation.' The results were as follows:

Progesterone	3.8 nmol/l
17β oestradiol	150 pmol/l
FSH	3 U/l
LH	2 U/l
Prolactin	1450 U/l

Why were the samples taken at this time, and what do the results mean?

10Q

This set of results came from a patient in chronic renal failure:

Na	K	Cl	CO_2	Urea	Creatinine
138	4.7	101	15	24.3	1082
mmol/l	mmol/l	mmol/l	mmol/l	mmol/l	μmol/l

Comment on the significance of the CO_2 and Cl results.

9A

Anovulation due to hyperprolactinaemia

The marked elevation in oestrogens and (particularly) progesterone which occurs in the luteal phase of the menstrual cycle can be used as a biochemical marker of ovulation. This patient's levels are in the 'pre-ovulatory' range, meaning that ovulation has not occurred at least in this cycle. The FSH and LH levels are not raised, however, excluding ovarian failure. The prolactin level is very high and this could well be causing failure of ovulation.

Hyperprolactinaemia was confirmed by cannulated samples (in case of a 'stress' result), and causes were looked for (e.g. pituitary tumour, hypothyroidism, drugs, chest trauma etc). This patient had biochemical hypothyroidism, and thyroxine treatment restored thyroid function tests, serum prolactin and fertility to normal.

10A

Uraemic acidosis

The plasma CO_2 level is low and the plasma chloride normal. The anion gap (total cations Na + K, less the total anions Cl + CO_2) is thus elevated at 27 mmol/l (it is usually about 15 mmol/l). The cause here is metabolic acidosis due to retention of unmeasured anions (e.g. sulphate, phosphate etc)—a common accompaniment of severely impaired renal function.

11Q

A 60-year-old woman had a rectal carcinoma removed surgically. There was no evidence of spread, and her LFT's at that time were normal. At follow up 3 years later however, the following results were noted:

Bilirubin	20 µmol/l
AST	23 U/l
Alkaline phosphatase	820 U/l

What is the likely cause, and what further investigations are needed?

12Q

A 47-year-old woman presented with dry eyes and pains in the knees and elbows. Amongst the investigations, the urea and electrolytes were done:

Na	K	Cl	CO_2	Urea	Creatinine
131 mmol/l	3.1 mmol/l	109 mmol/l	10 mmol/l	6.5 mmol/l	87 µmol/l

Comment on these results.

11A

Hepatic metastases

A markedly raised alkaline phosphatase, with normal or only slightly raised bilirubin and AST, constitutes 'dissociated LFT's' and is suggestive of a space occupying lesion in the liver, such as cyst, abscess or tumour. The mechanism is obstruction of a biliary radicle, causing secretion of liver alkaline phosphatase from the obstructed duct epithelium, but little or no hepatocellular damage or obstruction to outflow of bile. It is also useful in this situation to 'type' the alkaline phosphatase by electrophoresis, or assay an alternative enzyme of hepatic origin such as γ GT.

In this case a boney origin of the alkaline phosphatase was excluded by electrophoresis, and radio-isotope liver scan showed hepatic metastases, with one large deposit near the porta hepatis.

12A

Renal tubular acidosis

The patient had Sjögrens syndrome. There is an association between this and Type I (distal) renal tubular acidosis. The low CO_2 value is characteristic of metabolic acidosis, as it is too low for a respiratory alkalosis. Hyperchloraemia (with normal anion gap) and hypokalaemia are also typical. The urinary pH does not fall below 5.5 even if ammonium chloride (0.1 g/kg) is given. Nephrocalcinosis and osteomalacia may result.

13Q

A patient with Crohn's disease had the following results:

Total protein	35 g/l
Albumin	15 /l
Calcium	1.71 mmol/l
Magnesium	0.64 mmol/l
Phosphate	0.95 mmol/l
Alkaline phosphatase	60 U/l

Comment.

14Q

The following are results of a short Synacthen Test and Insulin Stress Test performed on a 49-year-old man (tests were performed at 9 a.m. on separate days).

Synacthen (250µg I/M)

0 minutes	cortisol	133 nmol/l
30 minutes	cortisol	430 nmol/l

IST (0.1 u/kg I/V)

time	blood glucose	plasma cortisol
0 mins	4.2	275
30 mins	1.3	260
60 mins	1.9	325
90 mins	2.1	330
120 mins	2.3	295

How would you interpret these?

13A

Hypoalbuminaemia

The only significant result is the low albumin level which may be due to malabsorption, or possibly hypercatabolism if the patient is postoperative or extremely ill. The low total Ca and Mg are due to reduced protein-bound fractions because of the small amount of available albumin.

There are numerous correction formulae for serum calciums—all of which are very much rules of thumb. A reasonable one is to add 0.1 mmol/l to the actual Ca result for every 6 g/l the albumin is below 40 g/l.

14A

ACTH deficiency

The adrenals respond to Synacthen, but the basal level is low for 9 a.m. This suggests a pituitary or hypothalamic disturbance. The insulin stress test achieves adequate hypoglycaemia, but the low basal levels of cortisol do not significantly increase, confirming a pituitary/hypothalamic disorder.

This patient was being investigated for obscure hyponatraemia. He had 'secondary' hypoadrenalism due to selective ACTH deficiency. Eventually, a pituitary tumour was demonstrated and removed, but the man required life-long cortisol replacement therapy.

15Q

Due to respiratory complications, a middle-aged woman was admitted to an intensive care unit for ventilation after an abdominal operation. On the 3rd post-operative day the blood gases were as follows:

pH	7.56
pCO_2	4.4 kPa
Base excess	+9 mmol/l
Actual bicarbonate	30 mmol/l
Standard bicarbonate	32 mmol/l
pO_2	12.9 kPa

What acid/base disturbance is present, and suggest a likely cause?

16Q

A middle-aged man with a family history of ischaemic heart disease had fasting lipid studies performed, with the following results:

Cholesterol	5.3 mmol/l
Triglycerides	3.0 mmol/l
Lipoprotein electrophoresis	Increased pre-beta fraction

What is the lipid disorder, and how would you proceed from here?

15A

Metabolic alkalosis

The high pH, base excess and bicarbonate levels with normal blood gases indicate an alkalaemia of metabolic origin. Examination of her fluid balance charts revealed nasogastric aspirates of over 2 litres per day since her operation. When a sample of the aspirate was tested it was found to have a pH of 3.5. This loss of hydrogen ion had thus caused a 'reduction' alkalosis.

16A

Type IV hyperlipoproteinaemia

Hypertriglyceridaemia with a raised pre-beta lipoprotein represents a Fredericksen Type IV abnormality. This may be secondary to other conditions such as alcoholism, obesity, diabetes, pancreatitis, chronic renal failure, acromegaly and gout. It may also occur as a familial disease. Treatment is generally aimed at the underlying disease. In this case the patient was obviously overweight, and lipids fell to normal on successful dieting.

17Q

A 50-year-old woman became acutely confused and soon lapsed into coma. Many investigations were performed with negative results, and eventually endocrine causes were explored. Thyroid function tests revealed the following:

                Thyroxine   25 nmol/l
                TSH   'undetectable'

What endocrine abnormality is suggested and how would you elucidate the problem?

18Q

These results were from a patient attending an ophthalmological follow-up clinic.

Na	K	Cl	CO_2	Urea	Creatinine
137	4.8	110	15	6.4	115
mmol/l	mmol/l	mmol/l	mmol/l	mmol/l	μmol/l

What is the abnormality and what is the likely cause?

17A

Secondary hypothyroidism

The slightly low serum thyroxine with undetectable TSH raises the possibility of hypothyroidism secondary to hypopituitarism. The thyroxine was repeated with a T_3 resin uptake test (to compute the Free Thyroxine Index). These still suggested mild thyroid underactivity, and a TRH test was done, showing a late rise in TSH 60 minutes after TRH administration—confirming the likelihood of primary hypothalamic or pituitary disease. Further evaluation of pituitary function—in particular cortisol levels—was precluded by the patient's demise. Autopsy revealed a diffuse viral meningo-encephalitis.

18A

Acetozolamide treatment

The abnormality is a hyperchloraemic acidosis, probably resulting from acetozolamide treatment for glaucoma. This drug inhibits carbonic anhydrase activity in renal rubular cells, preventing efficient hydrogen ion excretion.

19Q

A 69-year-old man was admitted with an acutely ischaemic right leg, thought to be due to a femoral artery thrombosis. His ECG was also abnormal with slight ST elevation and T wave inversion over the antero-septal leads. Three days later the house physician presented the following enzyme results to the consultant, suggesting that the patient had also had a silent MI.

	Day 1	Day 2
CK	1508	1230 U/l
AST	30	26 U/l

Comment!

20Q

A 65-year-old man developed jaundice associated with weight loss but no pain. Physical examination was normal apart from deep icterus. The liver function tests were as follows:

Bilirubin	223 μmol/l
AST	25 U/l
Alkaline phosphatase	570 U/l

What type of jaundice is this, and what is the likely diagnosis?

19A

Enzyme rises secondary to tissue ischaemia

Ischaemic muscle will release large amounts of various enzymes into the circulation—especially CK, and the diagnosis of myocardial infarction on the basis of enzyme changes is dangerous in these circumstances. CK isoenzyme electrophoresis may be useful, if available, to distinguish the brain ('BB'), skeletal muscle ('MM') and cardiac muscle ('MB') isoenzymes.

20A

Obstructive jaundice

The LFT's are clearly obstructive with marked elevations in bilirubin and alkaline phosphatase, but little change in aminotransferases. In caucasians of this age, obstructive jaundice is nearly always due to gall stones or carcinomas. With a history of weight loss, the latter is the most likely, and this patient turned out to have a carcinoma of the head of the pancreas.

Clinically, deep jaundice is usually obstructive and the diagnosis in this case was correctly made before biochemical results were available.

21Q

Glycosylated haemoglobin was estimated by electrophoresis in a black diabetic patient. The result was:

HbA_1 6.6%, with 'double peak' noted on electrophoresis

How would you investigate this?

22Q

A pregnant woman with a poor obstetric history had the following urea and electrolyte results:

Na	K	Cl	CO_2	Urea	Creatinine
141	3.7	108	18	4.3	55
mmol/l	mmol/	mmol/l	mmol/l	mmol/l	µmol/l

Comment.

21A

Haemoglobinopathy

Abnormal haemoglobin may be present. In this patient total Hb was 12.7 g/dl with numerous target cells present in the peripheral blood. Sickle test was negative and glucose 6-phosphate de-hydrogenase level was normal. Haemoglobin electrophoresis showed 58% HbA and 42% HbC.

22A

Hyperventilation of pregnancy

The low CO_2 could be due to metabolic acidosis or respiratory alkalosis. Blood gas analysis showed the latter to be the case. This is due to the slight hyperventilation which accompanies normal pregnancy.

23Q

A 58-year-old man was admitted with an exacerbation of chronic bronchitis. The following biochemical abnormalities were noted, and they persisted despite recovery from the infection:

Na	K	Cl	CO_2	Urea	Creatinine
116	3.0	74	20	2.4	54
mmol/l	mmol/l	mmol/l	mmol/l	mmol/l	µmol/l

What is the likely diagnosis, how would you confirm it, and what treatments may be useful?

23A

Bronchogenic carcinoma with 'ectopic ADH' syndrome

The general picture is of dilutional hyponatraemia, with marked hyponatraemia and reduction in concentration of other measured variables. This may be due to so-called 'SIADH' (syndrome of inappropriate ADH secretion) which complicates a variety of intrathoracic conditions, including straightforward bronchitis. The mechanism is thought to be due to reduced venous return stimulating atrial volume receptors, thereby causing ADH release (this is, however, speculative). This man's electrolyte abnormality persisted after recovery from the chest infection raising the possibility of an associated bronchogenic carcinoma with 'ectopic ADH' production. This was confirmed by sputum cytology and bronchoscopy, and was found to be inoperable.

'SIADH' tends to be over-diagnosed. Hyponatraemia, hypo-osmolality and inappropriately concentrated urine must be demonstrated, and in addition there must be normal renal and adrenal function; and no evidence of oedema, dehydration, hypo-volaemia or heart failure.

When SIADH is due to lung cancer, the water retention may cause no symptoms, but sometimes lethargy, malaise or confusion are troublesome. In these cases, and when the tumour cannot be removed, treatments such as water restriction, diuretics or demeclocycline may be useful. The latter is a tetracycline analogue which antagonises ADH action at the renal tubular level, and is a particularly useful treatment.

24Q

A longstanding epileptic who had also drunk alcohol excessively for several years was admitted for control of his seizures. The following investigations were returned:

Albumin	30 g/l	Bilirubin	15 μmol/l
Calcium	1.85 mmol/l	AST	15 U/l
Phosphate	0.6 mmol/l	Alkaline phosphatase	638 U/l

What are the possible causes for these abnormalities, and what further investigations would you consider?

25Q

A 23-year-old girl was admitted in a drowsy and confused state, having been well the previous day. Full blood count, chest X-ray and CSF examination were all normal, but urea and electrolytes were slightly abnormal:

Na	K	Cl	CO_2	Urea	Creatinine
135 mmol/l	4.0 mmol/l	97 mmol/l	14 mmol/l	6.5 mmol/l	212 μmol/l

What is the diagnosis?

24A

Anti-convulsant osteomalacia

As they stand these results raise possibilities such as osteomalacia (nutritional or due to anti-convulsants), cirrhosis or a space occupying liver lesion. The alkaline phosphatase must be 'typed', and in this case it proved to be of bone origin. Dietary assessment and tests of malabsorption were normal, as was a skeletal survey. However, bone biopsy confirmed osteomalacia which was assumed to be due to anti-convulsants.

25A

Diabetic ketoacidosis

The important abnormality is a high anion gap of 28 mmol/l due to a low CO_2 with normal Cl. This suggests metabolic acidosis with high levels of unmeasured anions. Diabetic ketoacidosis was confirmed when blood glucose was measured and the urine tested for ketones (the glucose result was 35.0 mmol/l and the urine showed + + for ketones). It later transpired that the girl was a known diabetic who was suffering from an acute urinary tract infection.

A further clue in this case is the raised creatinine with normal urea. Ketone bodies interfere with the common laboratory method (Jaffe reaction) for creatinine determination, and may give a spuriously high result.

26Q

A 57-year-old woman with longstanding severe chronic bronchitis and asthma was admitted in respiratory failure. Her blood gases were as follows:

pH	7.03
pCO_2	10.5 kPa
Base excess	−13 mmol/l
Standard bicarbonate	15 mmol/l
Actual bicarbonate	21 mmol/l
pO_2	6.1 kPa

What is the acid/base disturbance persent and what are the mechanisms involved?

27Q

A 64-year-old lady was found in a collapsed state at home. She had a history of ischaemic heart disease, and also took 10 mg prednisone daily for rheumatoid arthritis. She was referred to hospital where the admitting registrar considered a hypoadrenal crisis as a likely diagnosis. He took blood for plasma cortisol estimation and then gave intravenous hydrocortisone. She appeared to improve somewhat, but the doctor was rather surprised when the following result was returned the next day:

    Plasma cortisol   4834 nmol/l

What is going on?

26A

Mixed type acidosis

The low pH with hypoxia and hypercapnia indicates respiratory acidosis, but the considerable base deficit and low standard bicarbonate indicates a metabolic component. The most likely explanation is a Type A lactic acidosis due to tissue hypoxia and muscular fatigue.

27A

Drug interference with cortisol assay

Many laboratories measure plasma cortisol by fluorimetric methods, which are susceptible to drug interference. Spirinolactone is the usual culprit, and in this case a phone call to the patient's general practitioner confirmed that she was indeed on this drug. The original sample was assayed for cortisol by a protein binding method and the result was normal. The patient was subsequently shown to have had a myocardial infarction. Other drugs which may interfere with the fluorimetric cortisol assay are frusemide, fucidic acid and mepacrine.

28Q

A 30-year-old woman was admitted to hospital unconscious. She was hyperventilating and had a history of high alcohol intake in the past, but there were no other contributory findings. The following are the results of initial investigations:

pH	6.76	Na	135 mmol/l	Blood glucose	4.6 mmol/l
pCO_2	1.9 kPa	K	5.9 mmol/l		
Base excess		Cl	102 mmol/l	Urine ketones	negative
Standard bicarbonate	'off-scale'	CO_2	2 mmol/l		
Actual bicarbonate		Urea	3.4 mmol/l		
		Creatinine	133 μmol/l	Blood alcohol	110 mg/l
pO_2	14.8 kPa				

What further test would you request to confirm the diagnosis?

29Q

A 25-year-old man was admitted with acute abdominal pain. He gave a history of recurrent attacks of similar pain since childhood, and had had one negative laparotomy. The laboratory reported difficulty with some of their assays because of very milky serum, and so fasting lipid studies were done:

Cholesterol   9.5 mmol/l
Triglycerides   23.6 mmol/l
Lipoprotein electrophoresis—chylomicrons greatly increased

What is the diagnosis, and what other features of the condition may be present?

28A

Lactic acidosis

The blood gases and electrolytes show a very severe metabolic acidosis, with an extremely high anion gap of 37 mmol/l. The normal blood glucose and negative urine makes ketoacidosis unlikely, leaving lactic acid as a likely contender. The lactate level was in fact 19.3 mmol/l. The patient died, and at autopsy no permanent liver damage was found. The cause of death was thought to be Type B lactic acidosis due to acute alcohol ingestion rather than pre-existing liver disease.

Other causes of 'high anion gap acidosis' include renal failure, and poisoning with methanol, salicylate, ethylene glycol and paraldehyde.

29A

Type I hyperlipoproteinaemia

Massive chylomicronaemia with milky-white fasting serum and normal or only slightly elevated cholesterol indicates a Fredericksen Type I hyperlipoproteinaemia. Recurrent abdominal pain is a feature of this condition, which may be inherited as an autosomal recessive. Other characteristics are lipaemia retinalis, eruptive xanthomata and pseudohyponatraemia (the latter two were present in this patient). The condition is due to impaired uptake of triglycerides by peripheral tissues, due to either lipoprotein lipase deficiency or apoprotein C II deficiency.

30Q

An old man was found in a confused state in his flat where he lived alone. The admitting doctor found him to be hypothermic (rectal temperature 32°C) and in congestive heart failure. An ECG showed anterior T wave inversion only. Blood was taken for a number of tests, and the patient was then admitted for rewarming and diuretic therapy. The results returned as follows:

Na	130 mmol/l	pH	7.29	CK	40 U/l
K	4.0 mmol/l	pCO_2	3.0 kPa	AST	19 U/l
Cl	90 mmol/l	Base excess	−7 mmol/l	LDH	426 U/l
CO_2	18 mmol/l	Actual bicarbonate	19 mmol/l	LDH	electrophoresis —increase in LDH_1 and LDH_5
Urea	15.5 mmol/l				
Creatinine	138 µmol/l	Standard bicarbonate	22 mmol/l	Amylase	820 U/l
		pO_2	10.0 kPa		

Explain these results and suggest a diagnosis?

31Q

A 16-year-old 'brittle' diabetic was admitted to hospital routinely for stabilisation. She had checked her blood glucose by a home monitoring meter at 11 a.m. on the day of admission and the result was 8.6 mmol/l. Her admission blood glucose and plasma electrolytes (taken at 3 p.m.) were as follows:

Na	K	Cl	CO_2	Urea	Creatinine	Glucose
129 mmol/l	4.9 mmol/l	95 mmol/l	23 mmol/l	6.5 mmol/l	68 µmol/l	36.8 mmol/l

Her urine was free of ketones and she felt well throughout. What is the cause of the hyponatraemia?

30A

Late myocardial infarction with complications

The enzymes are consistent with a late MI with congestive heart failure (the CK and AST have returned to normal, and both liver and heart LDH isoenzymes are raised). The urea and electrolyte results are compatible with heart failure and also suggest a slight metabolic acidosis with a raised anion gap. The blood gases confirms the acidosis which is most likely to be a Type A lactic acidosis due to hypothermia (and possibly heart failure also). The raised amylase is probably due to hypothermia, and does not necessarily indicate clinically important pancreatitis.

The old man made a good recovery on appropriate therapy and admitted to an attack of severe chest pain 4 days prior to admission, after which he had retired to bed and not kept his flat heated.

31A

Hyponatraemia secondary to hyperosmolality

Any rapid and substantial rise in blood glucose concentration will also increase plasma osmolality (roughly by 1 mmol/kg for each 1 mmol/l glucose). In the diabetic, this glucose cannot be rapidly cleared into cells, and the extra-cellular hyperosmolality may be severe and prolonged. Equilibration across cell membranes results in water moving from cells into the ECF, causing hyponatraemia.

32Q

A 24-year-old man with a previous depressive history had an argument with his girlfriend and drank an unknown quantity of anti-freeze (ethylene glycol) in a suicidal attempt. He subsequently vomited several times, and when admitted to hospital 6 hours later, the following biochemical abnormalities were noted:

Na	K	Cl	CO_2	Urea	Creatinine		
150 mmol/l	4.3 mmol/l	109 mmol/l	22 mmol/l	5.6 mmol/l	100 µmol/l	pH	7.24
						pCO_2	5.6 kPa
						pO_2	11.4 kPa
						Actual bicarbonate	18 mmol/l
						Base excess	−9 mmol/l

Explain these abnormalities.

33Q

A 54-year-old woman developed complications following abdominal surgery, and was treated in an intensive care unit for respiratory failure and septicaemia. She required artificial ventilation. The following electrolytes and blood gases were performed in the lab one day:

Na	K	Cl	CO_2	Urea	Creatinine		
140 mmol/l	3.4 mmol/l	87 mmol/l	34 mmol/l	13.4 mmol/l	125 µmol/l	pH	7.68
						pCO_2	3.7 kPa
						BXS	+14 mmol/l
						Standard bicarbonate	38 mmol/l
						Actual bicarbonate	32 mmol/l
						pO_2	19.2 kPa

What do these results show, and suggest possible causes?

32A

Metabolic acidosis due to ethylene glycol

The blood gases shows an uncompensated metabolic acidosis, and though the plasma CO_2 is surprisingly well maintained, the anion gap is raised at 23 mmol/l (the hypernatraemia—probably related to vomiting—masks the 'acidotic electrolytes'). Ethylene glycol is metabolised to the acid radical oxalate, and leads to a metabolic addition acidosis with raised anion gap (like keto and lactic acidoses). Similar acid base disturbances may occur with excessive methanol or paraldehyde.

33A

Mixed alkalosis

The blood gases suggest over-vigorous artificial ventilation (High pO_2 and low pCO_2) causing a respiratory alkalosis. However, the raised standard bicarbonate and base excess also indicate a metabolic element. Possible causes in this case were hypokalaemia and recent bicarbonate therapy.

34Q

A 77-year-old woman was admitted severely ill with a lower respiratory infection and left ventricular failure. She was found to have glycosuria on routine ward analysis (2% Clinitest, negative for ketones). Plasma urea and electrolytes and blood glucose was requested:

Na	K	Cl	CO_2	Urea	Creatinine	Glucose
139	4.2	100	27	8.5	140	10.6
mmol/l	mmol/l	mmol/l	mmol/l	mmol/l	μmol/l	mmol/l

Is the patient diabetic, and how would you manage the hyperglycaemia?

35Q

A 35-year-old insulin dependent diabetic (of 14 years' duration) had suffered retinopathy and peripheral paraesthesia for some years. He had lately developed swelling of the ankles, and urine was positive for protein by 'Labstix' testing. His urea and electrolytes were as follows:

Na	K	Cl	CO_2	Urea	Creatinine
135	4.0	95	31	15.1	180
mmol/l	mmol/l	mmol/l	mmol/l	mmol/l	μmol/l

What other investigations are needed, and what is the likely diagnosis?

34A

'Stress diabetes'

Moderate hyperglycaemia and glycosuria without ketosis is a common accompanient to serious illness. It is due to high circulating levels of catabolic hormones such as cortisol, glucagon and catecholamines. Observation only is required providing that the blood glucose does not rise further. If blood glucose levels rise above 12 mmol/l short-term insulin therapy may be required. With antibiotics and diuretics this patient made a good recovery, and several days later a blood glucose 'series' was quite normal. In cases of doubt a formal glucose tolerance test should be performed 6 to 12 weeks later.

35A

Diabetic nephropathy

Serum albumin, urinary protein and creatinine clearance estimations are certainly needed, followed by renal biopsy if there is any doubt about the diagnosis. This patient's serum albumin was 25 g/l, total protein 55 g/l, creatinine clearance 53 ml/min, and urinary protein 9.5 g/24 h.

This nephrotic syndrome was later confirmed by biopsy to be due to diabetic glomerulosclerosis.

36Q

An 80-year-old woman was admitted in a drowsy and confused state at a weekend. No history was available other than that she lived alone and had not been seen by neighbours for some time. She was febrile (temperature 38.5°C) and her tongue was dry and dirty, but nothing else was found on examination. Admission urea and electrolytes were as follows:

Na	K	Cl	CO_2	Urea	Creatinine
163	3.4	126	28	17.8	140
mmol/l	mmol/l	mmol/l	mmol/l	mmol/l	μmol/l

What other immediate investigations would you like, and what is the likely cause of the electrolyte disturbance?

37Q

A sample was sent for urea and electrolyte measurement on a 75-year-old lady with an undiagnosed illness. Unexpected precipitation problems occurred in the Autoanalyser as the sample was processed, and an enthusiastic biochemistry registrar wondered if the patient may have a paraproteinaemia (which can cause such problems). He saved a sample for protein electrophoresis and was gratified when the result was as follows:

Total protein	66 g/l
Albumin	37 g/l
Electrophoresis	M band

However, the ward sent a repeat speciment and this showed no M band. Which result is right?

36A

Dehydration

The commonest cause of hypernatraemia is straight-forward dehydration due to loss of hypotonic body fluids, or failure to drink. The relatively greater increase in urea than creatinine is supportive of this (the old 'urea: creatinine ratio' is still useful in clinical medicine). The other immediate tests needed are a full blood count (likely to show a high haemoglobin and haematocrit) and blood glucose (to exclude hyperosmolar non-ketotic diabetic coma).

It eventually transpired that this woman had become generally unwell 5 days previously and had taken to her bed, and drunk and eaten very little. She was found to have a urinary tract infection, and when this and her dehydration was treated, she made a full recovery.

37A

Fibrinogen band

The registrar was wrong! The initial sample was for 'U & Es' and therefore a heparinised sample. When plasma is electrophoresed the fibrinogen fraction may simulate a myeloma band. The second sample was sent specifically for protein analysis (and was therefore clotted blood) and was normal, and thus the diagnosis was wrong.

38Q

A 31-year-old woman with end-stage cirrhosis was admitted to a surgical ward with an acute gastro-intestinal bleed and hepatic failure. Despite rescuscitation she did badly, and developed a chest infection. She eventually died 2 weeks after admission. For most of her admission she had disordered electrolytes, and at one stage blood gases were done:

Na	K	Cl	CO_2	Urea	Creatinine		
129 mmol/l	2.9 mmol/l	84 mmol/l	38 mmol/l	4.2 mmol/l	57 μmol/l	pH	7.56
						pO_2	7.2 kPa
						pCO_2	5.8 kPa
						Base excess	+16 mmol/l
						Standard bicarbonate	39 mmol/l
						Actual bicarbonate	38 mmol/l

Explain the disturbances.

39Q

A 4-week-old baby girl was severely ill following 2 operations for a duodenal atresia. The child was oedematous and jaundiced, and for some days had been noted to have abnormal 'U & Es' and also hypoalbuminaemia. The following results were eventually noted by the chemical pathologist:

Na	K	Cl	CO_2	Urea	Creatinine
137 mmol/l	3.9 mmol/l	112 mmol/l	24 mmol/l	1.1 mmol/l	29 μmol/l

Albumin 21 g/l

The pathologist phoned the ward, and was able to suggest a possible cause for the abnormal electrolytes.

Can you suggest what this was?

38A

Hypokalaemic alkalosis

The blood gases reveal a pure metabolic alkalosis with respiratory compensation. The likely cause from the electrolytes and clinical history is hypokalaemia. Potassium depletion in liver failure is usually multi-factorial, and factors involved include secondary hyperaldosteronism, low potassium intake and diuretic therapy. Hyponatraemia is also very common in liver disease, and again it is often difficult to put it down to one cause (possibilities include low salt intake—either voluntary or prescribed, diuretic therapy, sick cell syndrome, or sometimes secondary to potassium depletion).

39A

Low anion gap due to hypoalbuminaemia

Albumin exists partially in a positively charged ionic form at physiological pH, and when severe and prolonged hypoalbuminaemia occurs a compensatory fall in other cations or increase in anions may occur. The result is a low anion gap (in this case 5 mmol/l), which is a rare biochemical finding. The low plasma urea and creatinine concentrations in this child reflect low protein turnover due to reduced food intake.

ERRATUM

Page 40 under 39A line 2
for positively charged ionic form read negatively charged ionic form

40Q

A 59-year-old woman had been unwell for 1 month with malaise and dizziness, and had lately been vomiting. She had been weak and drowsy for 2 days and had collapsed in the street. When brought into the accident and emergency department she was unconscious and shocked. Urine was negative for glucose and ketones, blood count was normal and there was no history or evidence of head injury.

Plasma urea and electrolytes were as follows:

Na	K	Cl	CO_2	Urea	Creatinine
116	5.9	84	20	24.2	168
mmol/l	mmol/l	mmol/l	mmol/l	mmol/l	μmol/l

How would you manage the situation?

41Q

A 55-year-old man with psoriasis had serum proteins and electrophoresis measured as part of a research project. The clinicians were surprised when the result came back as follows:

Total protein	67 g/l
Albumin	38 g/l
Electrophoresis	M band

It was suggested that a biochemical error may have occurred because there was no clinical evidence of myeloma, and the protein levels were normal.

Is this likely, and what should be done?

40A

Addisonian crisis

The clinical and biochemical findings are strongly suggestive of hypocortisolaemia. A specimen of plasma or serum should be taken for cortisol determination (an urgent estimation is not required as the treatment will not be harmful), and hydrocortisone and saline given on the assumption that the patient is in an Addisonian crisis. In this case the result later came back as 30 nmol/l (the time of the specimen was 3 p.m.), and the patient made a remarkable recovery. She was later converted to prednisone (this steroid is not measured in most cortisol assays), and a Synacthen test confirmed the diagnosis of hypoadrenalism. Unfortunately, subsequent investigations revealed that the cause was disseminated ovarian cancer, with deposits in both adrenal glands.

41A

Myeloma with normal total globulins

It is dangerous to doubt a diagnosis of myeloma if an M band is found but the total globulin is normal. This patient was shown to have a monoclonal rise in IgG on immunoelectrophoresis, and the diagnosis was confirmed histologically on marrow examination. The first step, however, is to ask for a repeat specimen. As a general rule in biochemistry, important clinical decisions should not be made on the basis of single abnormal results—and in this case it is important to make sure that the electrophoresis was not done on plasma, which could give a 'false M band' due to fibrinogen (see 37Q).

42Q

An 18-month-old boy was 'failing to thrive'. There was no apparent cause, and the child was admitted to a paediatric ward for tests of thyroid and pancreatic function. The following were returned after the child left:

Na	K	Cl	CO_2	Urea	Creatinine
139	4.5	103	15	7.0	61
mmol/l	mmol/l	mmol/l	mmol/l	mmol/l	μmol/l

What diagnosis must be immediately excluded, and what other possibilities are there?

43Q

A 54-year-old woman with known asthma was admitted because of acute chest pain. It was noted that she had a myxoedematous appearance. Her ECG showed anterior ST depression and T wave inversion which did not alter. The following results were reviewed one week later:

	Day 1	Day 2	Day 3	Day 7	
CK	1243	904	1082	1120	U/l
T_4	20	—	—	—	nmol/l
TSH	>60	—	—	—	mU/l

Explain the CK results.

42A

Diabetes mellitus

A mild to moderate metabolic acidosis is suggested by the electrolytes, and, in the absence of renal impairment, diabetes is the important diagnosis to remember. This proved to be the case here and the child subsequently grew normally and showed normal development when treated with insulin.

Other less likely possibilities are one of the rare organic acidaemia of childhood, a Type B lactic acidosis (e.g. due to liver disease, malignancy or inborn metabolic errors) or poisons (such as ethylene glycol, methanol or salicylate).

43A

Myxoedematous myopathy

The persistent high elevation of CK is not consistent with a myocardial infarct, and hypothyroidism should be thought of as the likely cause. Raised CK levels in myxoedema may occur without a clinically obvious associated myopathy, though the source is still skeletal muscle. In this patient the CK levels dropped as her thyroid function tests returned to normal on thyroxine replacement.

44Q

Comment on the plasma urea concentrations in the following patients:

	Na mmol/l	K mmol/l	Cl mmol/l	CO_2 mmol/l	Urea mmol/l	Creatinine μmol/l
a. 28-year-old with alcoholic hepatitis	128	3.2	100	17	0.3	44
b. 24-year-old, 7 months pregnant	138	3.5	106	19	2.0	52
c. 3-year-old boy with nephrotic syndrome	139	3.3	99	26	0.6	31

45Q

The following results were phoned urgently to the ward from the biochemistry laboratory:

Na	K	Cl	CO_2	Urea	Creatinine
139	8.9	101	27	5.2	70
mmol/l	mmol/l	mmol/l	mmol/	mmol/	μmol/l

The specimen was not haemolysed and the potassium result was checked. When the ward was phoned there appeared to be no artefactual reason for the result (drip arm venepuncture, wrong tube etc.), but the patient was reported to have prolymphocytic leukaemia and had just started treatment.

What is the mechanism of the hyperkalaemia?

44A

'Hypouraemia'—varying causes

1. The electrolytes suggest that this patient has severely disturbed liver function—possibly liver failure—with hyponatraemia, hypokalaemia and metabolic acidosis. In this situation, plasma urea depression reflects reduced ability of the liver to synthesise urea from nitrogenous precursors.

2. Increased urea clearance and water retention are normal occurrences in pregnancy, and causes a lowering of plasma urea. Note also the high Cl and low CO_2 (hyperventilation of pregnancy).

3. In this child the low urea is likely to be due to severe urinary protein loss, resulting in protein deficiency and reduction in urea production.

45A

'Pseudohyperkalaemia' due to cell breakdown

Intracellular K levels are high (about 160 mmol/l), and white cell destruction may cause hyperkalaemia (as from red cells when a blood sample is haemolysed). The raised K levels from leukaemic white cells is usually an in vitro effect, with the K ions escaping from leucocytes in the blood tube after the sample has been taken—hence the term 'pseudohyperkalaemia'. In this patient a repeat sample was rapidly centrifuged after venepuncture and the plasma K was 5.3 mmol/l. In some patients, however, marked hyperkalaemia persists despite these precautions—presumably due to in vivo K release from leukaemic white cells.

46Q

A 42-year-old man was admitted with renal colic, and subsequently had blood tests for serum calcium levels performed. His 'bone chemistry' was as follows:

Calcium	2.74 mmol/l
Phosphate	1.18 mmol/l
Albumin	54 g/l
Alkaline phosphatase	71 U/l

On the basis of these he was investigated for primary hyperparathyroidism. Do you think the diagnosis was comfirmed or not, and give reasons?

47Q

A registrar in clinical pharmocology was surprised at the following U & E result on a fit 30-year-old male who was taking part in a research project:

Na	K	Cl	CO_2	Urea	Creatinine
139	4.2	101	25	25.2	64
mmol/l	mmol/l	mmol/l	mmol/l	mmol/l	μmol/l

He sent a further specimen, and on this one the urea concentration was normal.

What might have gone wrong?

46A

Hypercalcaemia with hyperalbuminaemia

The normal phosphate is not supportive of primary hyperparathyroidism, and if the calcium is 'corrected' for the slightly high albumin level it is brought into the normal range. It turned out that the sample was taken after breakfast, and the patient had vomited several times the previous day. Further calcium and PTH levels were normal.

There is considerable controversy over the validity of 'correction formulae' for total calcium, particularly in hyperalbuminaemic states. Nevertheless, they provide a useful guide to clinicians as long as it is remembered that they are very much approximations. Any attempt to evaluate serum calcium levels should thus involve 3 specimens taken fasting on a normally hydrated patient, without using a tourniquet. If there is an albumin abnormality and the 'corrected' level is at all borderline, then further investigation should ensue. Direct measurement of ionised calcium is now available in many laboratories.

47A

Print-out error

The elevation in urea alone is most odd, and when the problem came to light, the original sample was re-analysed and the urea level was normal. Autoanalysers are rarely wrong, but occasional random misprints occur at the recording end. Such problems are usually picked up by the laboratory staff, and the sample repeated. This one got through the net. If urea and creatinine are measured on separate analysers, transposition of either specimen or result should be considered.

48Q

A GP requested an urgent out-patient appointment with a chest physician because of a suspicious lung shadow on a CXR performed because of haemoptysis. The physician admitted the patient directly as he was unwell, and asked the house physician to perform blood gases, blood count, ESR, urine analysis, urea and electrolytes. The first results were as follows:

Blood gases	pH	7.56	Urinalysis
	pO_2	7.5 kPa	glucose—2%
	pCO_2	6.8 kPa	ketones—negative
	Actual bicarbonate	45 mmol/l	
	Base excess	+21 mmol/l	

What might the urea and electrolytes have shown, and what diagnosis do you suspect?

49Q

A 15-year-old girl with known portal hypertension (due to a previous portal vein thrombosis) was admitted with bleeding oesophageal varices. Her U & Es the next day were as follows:

Na	K	Cl	CO_2	Urea	Creatinine
140	3.8	106	23	12.6	84
mmol/l	mmol/l	mmol/l	mmol/l	mmol/l	μmol/l

What does the elevated urea signify?

48A

'Ectopic ACTH' syndrome

The U & E showed a marked hypokalaemic alkalosis with a plasma K of 2.3 and CO_2 of > 40 mmol/l. Blood glucose was 19.6 mmol/l and diurnal cortisols showed levels of 3120 nmol/l at 9 a.m. and 3500 at 12 MN. Subsequent bronchoscopy and biopsy confirmed an oat cell tumour which was inoperable, and ACTH levels were later shown to be grossly elevated. On close examination the patient had buccal hyperpigmentation and later in his admission became generally hyperpigmented. He achieved metabolic normality on a carbohydrate controlled diet, glibenclamide, metyrapone and dexamethasone, He died 6 months later.

49A

Uraemia due to gastrointestinal haemorrhage

A mild to moderate elevation in plasma urea is almost universally seen for a few days after an acute GI bleed. The standard explanation is increased absorption of nitrogenous material, but recently this has been doubted and it may be a direct renal effect.

50Q

A 31-year-old woman saw her general practitioner because of mild galactorrhoea. Menstruation was normal, she had breast fed her 2 children with no problems—the last being born several years ago. The doctor performed a number of tests including the following:

Prolactin	400 mU/l
Growth hormone	19.2 mU/l

Because of the high GH she was referred to an endocrinologist with a diagnosis of '? pituitary tumour'. The consultant found her to be clinically normal apart from very mild galactorrhoea to expression. He was able to suggest a cause for the hormonal abnormality and reassure both general practitioner and patient. What was it?

51Q

A 49-year-old lady was being investigated by an orthopaedic surgeon for easy fracturing. The first result back was as follows:

Na	K	Cl	CO_2	Urea	Creatinine
141	3.8	111	15	6.7	110
mmol/l	mmol/l	mmol/l	mmol/l	mmol/l	μmol/l

Is there any clue to the diagnosis here?

50A

Stress—induced rise in GH level

On questioning the patient, the bloods in the GP surgery has been taken by direct venepuncture. GH in particular (but also prolactin) is very susceptible to rises in response to such stress and this is almost certainly the explanation.

Full evaluation revealed this patient to have 'normoprolactinaemic galactorrhoea' which eventually settled spontaneously.

As a general rule, single blood measurements of pituitary hormones are of little use (with the exception, perhaps, of TSH). Many are secreted in a pulsatile fashion (e.g. ACTH), some can be elevated by venepuncture stress (e.g. GH and prolactin), and some vary greatly diurnally (e.g. ACTH) or menstrually (e.g. FSH and LH). Low or undetectable levels may be normal, and small elevations may not be real because of high assay variability. 'Dynamic' tests are therefore needed to evaluate pituitary function properly (e.g. insulin stress test, TRH and LRH tests etc.).

51A

Hyperparathyroidism

The high Cl and low CO_2 suggest a hyperchloraemic acidosis, and in the clinical situation this may well be due to CO_2 acquired renal tubular acidosis. This is thought to be due to a direct effect of high PTH levels on the renal tubules. It is surprisingly common and old textbooks mention that CL levels below 100 make the diagnosis doubtful.

This patient was found to have a Ca level of 3.08 mmol/l with normal albumin, P of 0.7 mmol/l and AP of 311 U/l.

52Q

A 28-year-old woman was 5 months pregnant. At an ante-natal visit she was noted to be jaundiced. This was thought to be 'cholestasis of pregnancy' and LFT's were performed. These returned as follows:

Bilirubin	238 μmol/l
AST	66 U/l
Alkaline phosphatase	350 U/l

What do you think?

53Q

A 19-year-old girl in a psychiatric hospital drank a quantity of surgical spirit with suicidal intent. She was admitted to a medical ward 4 hours later when her U & E's were as follows:

Na	K	Cl	CO_2	Urea	Creatinine	Glucose
145 mmol/l	4.8 mmol/l	91 mmol/l	5 mmol/l	9.0 mmol/l	120 μmol/l	9.0 mmol/l

What is the cause of the biochemical abnormality?

52A

Acute fatty liver of pregnancy

Although the AP rise is proportionately more than the AST, there is probably a combination of hepatocellular and obstructive elements. The degree of dysfunction is not compatible with simple cholestasis of pregnancy, and infective hepatitis or acute fatty liver of pregnancy is more likely. Biopsy proved the latter more sinister diagnosis to be the case. The patient subsequently became severely ill with worsened liver function and acute renal failure. She survived, but lost the baby.

53A

Metabolic acidosis due to methanol

The low CO_2 with normal Cl suggest a metabolic acidosis, and the anion gap computes to an exceedingly high 54 mmol/l. Methanol may cause such a metabolic acidosis by being metabolised to formic acid. Remember other causes of a high anion gap—ketoacidosis, lactic acidosis, salicylate poisoning, and overdose with ethylene glycol and paraldehyde.

54Q

A 34-year-old woman was referred with hirsutes. This was thought to be constitutional, but as it was causing her some distress, treatment was begun with cyproterone acetate. After a long delay, however, the following initial hormonal results returned (samples taken on the 5th day of her menstrual cycle).

FSH	9 U/l
LH	18 U/l
Prolactin	300 mU/l
Testosterone	5.1 nmol/l
Urinary 11 OHCS	240 nmol/l

What do you think?

55Q

A sample for U & E was received routinely and the form gave the following as the clinical details—'Cushings, on steroids and diuretics'. The patient was a 31-year-old female and the result was as follows:

Na	K	Cl	CO_2	Urea	Creatinine
165	3.6	97	22	5.2	74
mmol/l	mmol/l	mmol/l	mmol/l	mmol/l	µmol/l

The biochemistry senior registrar could think of no reason why hyperadrenalism, steroid treatment or diuretics could cause such hypernatraemia and phoned the ward. The problem was rapidly solved.

Any ideas?

54A

Polycystic ovary syndrome

This condition is complex and poorly understood, and there are no diagnostic biochemical features. However, the raised gonadotrophins (especially LH) and testosterone levels are suggestive though idiopathic hirsutes, and rarely adult presenting congenital adrenal hyperplasia or virilising tumour remain possibilities. This woman was also somewhat overweight and her periods were irregular—though these are of course 'soft' endocrine symptoms. She was referred gynaecologically and eventually the diagnosis was confirmed laparoscopically.

55A

Indwelling cannula flushed with sodium citrate

The patient was having a battery of blood tests including an insulin stress test, and an indwelling cannula had been inserted. The houseman was new to the ward and had used sodium citrate to keep the cannula patent prior to taking the U & E sample. A repeat specimen, properly taken, gave an Na value of 144 mmol/l.

The high sodium with normal Cl was the clue to something suspicious here. The anion gap was 49—and citrate can thus be added as a very rare cause of raised anion gap:

56Q

A 62-year-old lady had a parathyroidectomy for primary hyperparathyroidism and subsequently required vitamin D therapy. Her serum calcium levels had been difficult to control, and she had required various dosages of differing vitamin D preparations—including latterly 1α hydroxycholecalciferol as her Ca levels had been low. She was admitted in an ill and dehydrated condition having been confused and vomiting for 3 weeks previously. Biochemical results were as follows:

Calcium	4.27 mmol/l
Phosphate	1.07 mmol/l
Albumin	41 g/l
Alkaline phosphatase	70 U/l

What is the likely diagnosis?

57Q

A 20-year-old girl attended a rheumatology clinic with systemic lupus errythematosis, involving joints and kidney. She was being treated with prednisone, azothiaprine and frusemide and was symptomatically well, despite mild renal impairment. Six weeks after a holiday in Spain she developed jaundice for which she was admitted to hospital. She felt otherwise well and her LFTs were as follows:

Bilirubin	193 μmol/l
AST	93 U/l
Alkaline phosphatase	84 U/l

What possible diagnoses exist and which is most likely?

56A

Vitamin D intoxication

Severe hypercalcaemia is a danger to all patients on vitamin D, especially the potent 1 α derivative. Close examination of out-patient records in this patient showed that her doses had been gradually increased (because of low Ca levels) until she was on 2 μg 1 α OH cholecalciferol, 6 g calcium gluconate and 180 μg vitamin D (the other 2 in a combined preparation). It is likely that the patient was not taking treatment properly, but started doing so shortly before her presenting illness.

57A

Drug-induced jaundice

The possibilities here are very wide and include viral hepatitis, 'opportunist' infections due to the immunosuppressive therapy, cholelithiasis, lymphoma, drugs and haemolysis. In a patient such as this, drug-induced jaundice is the most probable cause.

Infective and obstructive causes were excluded by serological and ultrasound studies. A liver biopsy revealed a predominantly cholestatic picture, and the jaundice cleared when the azothiaprine was stopped (this drug can also cause hepatocellular damage). The LFTs show the interpretive difficulties which occur not infrequently in clinical practice. The bilirubin is surprisingly high for hepatocellular jaundice, yet the enzymes pattern is not 'obstructive'; it is important to recognise the limitations of biochemistry in this situation—liver biopsy is the investigation of choice, after an ultrasound scan for obstructive causes.

58Q

A 20-year-old obese girl was referred to a diabetic clinic because of glycosuria and vaginal candidiasis. She had a family history of diabetes, but she had no weight loss, thirst or polyuria. When she attended her urine contained a 'trace' of glucose but no ketones, and a random blood glucose was 7.9 mmol/l. She was put on a weight reduction diet and an out-patient glucose tolerance test was arranged for the next week. The result was as follows:

	Time	Blood glucose (venous whole blood)	Urine glucose	Urine ketones
75 g glucose →	0 minutes	7.4	trace	negative
	60 minutes	12.3	½%	negative
	120 minutes	12.0	2%	negative

What is the diagnosis?

59Q

A 10-year-old girl was severely asthmatic and was admitted one night in severe status asthmaticus, having been deteriorating for several days previously. Due to difficulty in performing an arterial puncture, her venous U & E returned first:

Na	K	Cl	CO_2	Urea	Creatinine
135 mmol/l	5.2 mmol/l	96 mmol/l	33 mmol/l	2.9 mmol/l	43 μmol/l (specimen not haemolysed)

Any comments?

58A

Maturity-onset diabetes of the young (MODY)

The story and findings are not typical of Type 1 (insulin dependent) diabetes, and other possibilities are renal glycosuria or MODY diabetes (maturity onset diabetes of the young—a dominantly inherited form of diabetes whose name describes exactly what it is). The GTT is diagnostic of diabetes (fasting venous blood glucose over 7.0 mmol/l and/or 2 hour post 75 g glucose load level of over 10.0 mmol/l), and the girl therefore has MODY diabetes.

A possible diagnostic pitfall here is that the patient was put onto a diet before the GTT. It is well recognised that fasting, or low carbohydrate diets, can induce glucose intolerance in nondiabetics. However, if this is the case, the fasting blood glucose is usually low.

59A

CO_2 retention—situation grave

Asthma which has decompensated to the stage of CO_2 retention is extremely severe, and the raised CO_2 in this girl's electrolytes is ominous, as is the mild hyperkalaemia which may be secondary to acidosis. Note the rather low urea and creatinine which could be related to poor food intake recently—another significant indication of the severity of her illness.

Her blood gases in fact showed a pH 6.85, CO_2 6.6 kPa, pO_2 9.1 kPa, base excess-13 mmol/l, standard HCO_3 14 mmol/l and actual HCO_3 31 mmol/l, i.e. very severe respiratory acidosis with a probable lactic acidosis related to extreme muscular effort and tissue hypoxia. The girl needed paralysis and ventilation as well as bronchodilator and steroid therapy, but she eventually recovered.

60Q

An 84-year-old man who lived alone had been progressively weak and ill for some months. He became unable to leave his upstairs flat and was eventually admitted to hospital. Apart from generalised weakness and inability to walk unaided, examination was unrewarding. A battery of investigations were done, amongst which were the following:

Calcium	1.61 mmol/l
Phosphate	0.73 mmol/l
Alkaline phosphatase	500 U/l
Albumin	42 g/l

Diagnosis and treatment?

61Q

An elderly man was referred to a medical out-patient clinic because of difficulty in walking and a variety of other complaints. He had been taking frusemide with potassium supplements for about a year, originally given for ankle oedema. U & E results were as follows:

Na	K	Cl	CO_2	Urea	Creatinine
132	2.5	89	30	9.0	129
mmol/l	mmol/l	mmol/l	mmol/l	mmol/l	μmol/l

Comments?

60A

Osteomalacia

Dietary assessment showed a very low vitamin D intake, and he rarely saw sunlight. X-rays revealed several Looser's zones. He returned to full mobility with vitamin D therapy, and regained independence. Though malabsorption and hypoparathyroidism must be borne in mind, in the elderly nutritional osteomalacia is by far the commonest cause of hypocalcaemia. Its presentation is rarely with waddling gaits, tetany or fractures—but more commonly with non-specific deterioration, as in this case.

61A

Disordered electrolytes due to diuretics

Every parameter is abnormal, and all can be explained by diuretics (hypokalaemia with alkalosis, hyponatraemia, hypochloraemia, and reduced renal function—though note that the reference range for urea and creatinine is a little higher in the elderly). With or without potassium supplements, diuretics are the commonest cause of disordered electrolytes. The patient did well once the drugs were stopped. Further biochemical problems with diuretics are glucose intolerance and elevations in serum urate.

62Q

A 28-year-old man was investigated for infertility and was found to be severely oligospermic. The following endocrine results were returned:

FSH	20.4 U/l
LH	18.6 U/l
Testosterone	3.2 nmol/l
Prolactin	350 mU/l

What do these indicate and what is the prognosis?

63Q

A 36-year-old man required prolonged ventilation in an intensive care unit for severe Guillain-Barré syndrome. He became persistently hyponatraemic, and the following investigations were performed:

Na	K	Cl	CO_2	Urea	Creatinine	Glucose
124 mmol/l	4.0 mmol/l	90 mmol/l	22 mmol/l	4.5 mmol/l	51 μmol/l	6.2 mmol/l

Plasma osmolality   257 mmol/kg
Urine osmolality   385 mmol/kg

Creatinine clearance, Synacthen test and serum lipids were also normal.

What may be the cause of the hyponatraemia?

62A

Primary testicular failure

The high gonadotrophin levels with low testosterone suggest primary testicular disease with elevated FSH and LH levels by negative feedback. The cause may be absence of functioning testicles (e.g. cryptorchidism) or any active or past destructive lesions. With such elevated gonadotrophin levels, the outlook for improving spermatogenesis is always poor.

63A

'Inappropriate ADH secretion'

The results are consistent with 'SIADH' which is well described in the Guillain-Barré syndrome. It is, however, probably due to *appropriate* ADH secretion in response to hypovolaemia (pooling of blood in the limbs and reduced pressure in the right atrium and great veins, where volume receptors are situated). Chronic hypovolaemia may eventually cause 'resetting' of hypothalamic osmoreceptors.

64Q

A 45-year-old man was severely ill with viral encephalitis, and required ventilation. After several days the consultant in charge noted him to be slightly hypokalaemic and to have some Cushingoid features. He asked for diurnal cortisol levels to be taken. The results were:

DAY 1	9 a.m.	1562 nmol/l
	12 MD	2230 nmol/l
DAY 2	9 a.m.	1840 nmol/l
	12 MD	1956 nmol/l

He asked for an endocrine opinion. What do you think?

65Q

A 35-year-old man with symptomatic chronic pancreatitis underwent surgery involving a partial pancreatectomy and drainage procedure. Recovery was complicated by development of a fistula along the track of a drain from the main pancreatic duct. The following electrolyte abnormality was also observed:

Na	K	Cl	CO_2	Urea	Creatinine
136	4.5	109	13	9.9	102
mmol/l	mmol/l	mmol/l	mmol/l	mmol/l	μmol/l

What is the explanation?

64A

Stress-induced hypercortisolaemia

The high cortisol levels with loss of diurnal rhythm are quite appropriate to the degree of stress involved with such a severe medical illness, ventilation, and all the other necessary medical and nursing procedures. In these situations plasma cortisol may be very high indeed. ACTH levels and the effect of dexamethasone suppression may help to elucidate the problem, but overactivity of the hypothalamic—pituitary—adrenal axis cannot be ideally investigated in such stressful situations.

65A

Hyperchloraemic acidosis due to bicarbonate loss

Pancreatic secretions are rich in bicarbonate, and when this is lost externally directly a metabolic 'subtraction' acidosis may occur, with a compensatory rise in Cl and normal anion gap i.e. a hyperchloraemic acidosis. Other causes of this type of acidosis are renal tubular dysfunction, acetozolamide treatment and ureteric implantation into the colon (due to Cl and CO_2 exchange across the colonic mucosa).

66Q

A middle-aged man was sent to casualty with acute chest pain. His ECG was equivocal, but he was nevertheless admitted to a coronary care unit with a presumptive diagnosis of myocardial infarction. The ECG did not alter over the next 3 days, and serial levels of cardiac enzymes are given below:

	Day 1	Day 2	Day 3	
CK	243	93	50	U/l
AST	15	17	12	U/l
LDH	178	150	180	U/l

Has the man had a myocardial infarction?

67Q

A 71-year-old patient with diabetes was coincidentally found to have a slightly raised serum calcium on one occasion. There consecutive fasting samples were then taken for calcium and PTH studies. The results were as follows:

	Ca	P	Albumin	AP	PTH
Day 1	2.70	0.75	38	70	2.0
Day 2	2.62	0.82	41	65	2.0
Day 3	2.71	0.69	39	72	1.8
	mmol/l	mmol/l	g/l	U/l	U/l

What is the diagnosis?

66A

Artefactually raised CK

Without rises in AST and LDH and with an unsupportive ECG, it is dangerous to diagnose an MI only on the basis of the CK. Enquiries in casualty showed that the patient had received an intramuscular injection of diamorphine and metoclopramide soon after admission. This was almost certainly the cause of the raised CK, and an MI cannot be diagnosed. The presence of the 'MB' isoenzyme of CK may be used to confirm myocardial origin.

67A

Primary hyperparathyroidism

The Ca levels are slightly but at least on 2 occasions definitely elevated. The low or low-normal P levels are consistent with hyperparathyroidism. The PTH levels are definitely raised—inappropriately so for such Ca levels, and confirming primary hyperparathyroidism. Many cases of 'borderline hypercalcaemia' turn out to be due to hyperparathyroidism on close investigation—the problem is whether such mild, asymptomatic cases, especially in the elderly, should be operated on.

68Q

A 41-year-old man was sent to a dermatologist because of skin lesions on the elbows and hands. These were diagnosed as palmar xanthomata and fasting lipids were arranged:

Cholesterol	14.4 mmol/l
Triglycerides	5.5 mmol/l
Electrophoresis	broad β band

What is the diagnosis?

69Q

A 48-year-old woman was referred to a physician because of possible hypothyroidism. She had developed some cold intolerance and dry skin, and had put on some weight recently. There was a family history of thyroid disease. The consultant's investigations returned as follows:

Thyroxine	74 nmol/l
TSH	13.9 mU/l

Thyroid auto-antibodies were also positive. What is your assessment of her thyroid status?

68A

Type III hyperlipoproteinaemia

The presentation and findings are typical of this unusual lipid disorder. This patient did extremely well on a weight reducing and fat restricted diet, and drugs were not needed.

The condition is due to the accumulation of cholesterol-rich VLDL (or intermediate density lipoprotein, IDL) which has an abnormal electrophoretic mobility. It is thought to be caused by apoprotein E III deficiency.

69A

Mild hypothyroidism

The most sensitive laboratory test of hypothyroidism is serum TSH, which is elevated in the early stages of hypothyroidism before serum thyroxine falls below the reference range. This woman presumably has autoimmune thyroiditis, and probably warrants thyroxine replacement at this stage. Early thyroid failure such as this is sometimes known as 'subclinical' or 'compensated' hypothyroidism.

70Q

A 69-year-old man was admitted with pneumonia. The admitting registrar was surprised at the following results:

Na	K	Cl	CO_2	Urea	Creatinine
134	4.4	68	22	19.3	142
mmol/l	mmol/l	mmol/l	mmol/l	mmol/l	μmol/l

What do you think?

71Q

A 50-year-old tramp was seen in outpatients with blisters on the hands and face. The following investigations were done:

plasma	Bilirubin	17 μmol/l
	AST	150 U/l
	Alkaline phosphatase	90 U/l
urinary	Porphobilinogen	14 μmol/24 hours
	Uroporphyrin	2840 nmol/23 hours
	Coproporphyrin	610 nmol/24 hours

What is the diagnosis?

70A

Laboratory mistake

The chloride results looks suspicious, and the test should be first repeated. 'Laboratory error' is in fact nowadays rare, though the lab is often blamed when results are difficult to interpret. In this case a second specimen gave a Cl level of 95 mmol/l, and despite extensive searches in the laboratory, no cause could be found for the original result.

71A

Porphyria cutanea tarda

This condition may occur in middle-aged men who have liver damage. It is associated with large increases in uroporphyrin excretion while increases in coproporphyrin are more modest. It should be differentiated from variegate porphyria in which neurological manifestations occur. Increased faecal protoporphyrin occurs in the latter condition. This patient was later found to be a methylated spirits addict.

72Q

A recently diagnosed 68-year-old diabetic who was overweight had fasting serum lipids measured:

    Cholesterol   9.0 mmol/l
    Triglycerides   2.2 mmol/l
    Electrophoresis—increased β and pre-β bands

What is the abnormality and what is it due to?

73Q

A 42-year-old man with vague complaints was strongly suspected of abusing alcohol. Liver function tests were as follows:

    Bilirubin   14 μmol/l
    AST   16 U/l
    Alkaline phosphatase   75 U/l

These were repeated on a number of occasions, with similar results. What other non-invasive tests may help to make the diagnosis?

72A

Type II B hyperlipoproteinaemia

The mild hypercholesterolaemia and hypertriglyceridaemia are likely to be due to the diabetes and obesity, and will improve when these are treated. It is doubtful if the performance of such tests is justified in this situation, as therapy will not be altered by the result.

73A

γ GT and ethanol levels

Random measurement of blood ethanol from samples taken in clinic may be positive. The enzyme gamma glutamyl transpeptidase (γ GT) may be elevated in significant alcohol abuse, when other liver enzymes are normal. In this case it was raised at 120 U/l, and subsequent liver biopsy confirmed the diagnosis of alcoholic liver disease. Other investigative parameters which are raised by excessive ethanol intake are the mean corpuscular volume (MCV), serum urate, and fasting serum triglyceride level.

74Q

A 35-year-old lady presented with a 12 month history of diarrhoea and fatigue. Investigations revealed the following:

Calcium	1.81 mmol/l
Albumin	32 g/l
Alkaline phosphatase	180 U/l (typed as 'bone')

What is the cause of these abnormalities, and what may be the underlying diagnosis?

75Q

An epileptic patient had 'routine' LFTs done at a yearly neurology clinic visit. The registrar who sent the test considered them normal, but a medical student on the ward at the time asked why the bilirubin level was low. The results were:

Bilirubin	1 μmol/l
AST	16 U/l
Alkaline phosphatase	70 U/l

The registrar was unable to provide an answer. Can you?

74A

Adult coeliac disease

The calcium level is low despite 'correction' for the hypoalbuminaemia, and this with the history and elevated alkaline phosphatase of bone origin suggests osteomalacia due to malabsorption. A Crosby Capsule biopsy confirmed the clinical suspicion of adult coeliac disease and the patient did well on a gluten free diet.

75A

Enzyme induction by barbiturates

The patient was on long term barbiturates, which are non-specific inducers of many hepatic enzymes. Breakdown of bilirubin is increased with consesquent low plasma levels. This is of no clinical significance, but perhaps less trouble would have been caused if the doubtfully necessary 'routine LFT's' had been omitted!

76Q

A 26-year-old diabetic presented to casualty with acute abdominal pain. The duty doctor arranged for urgent blood glucose, urea and electrolytes, blood count and serum amylase. The first results back were as follows:

Hb	15.0 g/dl
WCC	16 500 mm^{-3}
Amylase	900 U/l

On the basis of these the patient was sent to a surgical ward as a case of acute pancreatitis. Subsequent results, however, put this diagnosis in doubt. What do you think?

77Q

A grossly obese middle-aged man was admitted to hospital for a near starvation diet. He was closely monitored biochemically, and this included regular urate levels. His consultant was surprised at the results when he saw him on his ward round.

Day 7 of diet	urate	0.60 mmol/l
Day 10 of diet	urate	0.58 mmol/l
Day 14 of diet	urate	0.64 mmol/l
Day 15 of diet	urate	0.25 mmol/l
Day 21 of diet	urate	0.62 mmol/l

Can you suggest a possible explanation for the result on day 15?

76A

Diabetic ketoacidosis

This is a common mistake. Elevated amylase levels, abdominal pain, and leucocytosis are all consistent with diabetic ketoacidosis—a diagnosis which was confirmed when the blood glucose and electrolytes results returned.

Pancreatitis is not the only condition which causes elevated amylase levels. As well as diabetic ketoacidosis, acute abdominal conditions, renal failure, mumps and macroamylassaemia may cause modest elevation of the serum amylase. The WCC is always elevated in severe diabetic ketoacidosis and correlates with blood ketone body levels rather than with the presence of infection.

77A

Wrong person's result

The patient is running slightly raised urate levels consistent with the increased purine breakdown associated with such a diet. The level on Day 15 is very much lower than the rest and suggests a mistake. It eventually transpired that the result probably belonged to a patient in the next bed who was being investigated for polyarthritis, and resulted from a mix up of names on the blood tubes. When mistakes do occur with blood results, problems such as this are far more common than analytical errors—though clinical staff tend to blame the latter!

78Q

A 40-year-old man with a possible myocardial infarction was noted to have periorbital xanthelesmata. He also had a strong family history of premature ischaemic heart disease. A fasting lipid profile was requested with the following results:

Cholesterol	12.1 mmol/l
Triglycerides	1.4 mmol/l
Electrophoresis	marked increase in β fraction

What does this indicate?

79Q

A 78-year-old lady had suffered 'funny turns' for about a year. These were becoming more frequent, and she found that she could revive herself by taking food. She was admitted to hospital where blood samples were taken during 4 such episodes, with the following eventual results.

	Blood glucose	Serum insulin
Episode 1	2.0 mmol/l	18.0 mU/l
Episode 2	1.8 mmol/l	33.0 mU/l
Episode 3	1.8 mmol/l	17.5 mU/l
Episode 4	1.4 mmol/l	16.2 mU/l

What is the diagnosis?

78A

Type IIA hyperlipoproteinaemia

The hypercholesterolaemia with increase in β lipoprotein indicates a Type IIA abnormality in the Fredericksen classification. However, it should be mentioned that the stress of myocardial infarction affects lipid measurements, and valid results cannot be expected for about 2 months after the event. In this case, the MI was not confirmed and the lipid abnormality was real.

There are at least 2 forms of primary hypercholesterolaemia—familial hypercholesterolaemia which is inherited as an autosomal dominant, and polygenic hypercholesterolaemia. The Type IIA pattern may also occur in some patients with familial combined hyperlipidaemia.

79A

Insulinoma

The episodes are obviously due to hypoglycaemia, and the serum insulin levels are clearly inappropriately high. This is most probably due to autonomous endogenous insulin secretion i.e. insulinoma. A less likely possibility (in this case) is factitious hypoglycaemia due to self-administration of insulin. If necessary, this is best distinguished by measurement of C-peptide levels (a marker of endogenous insulin production) during hypoglycaemia. Because of this lady's age and general health, the surgeons were not happy to operate, but she subsequently did well on diazoxide therapy.

80Q

A 58-year-old man became unwell and developed jaundice. His general practitioner diagnosed infective hepatitis and arranged for some liver function tests to be done. After a week, however, the man was somewhat more icteric, and the general practitioner decided that the jaundice must be 'surgical' and arranged appropriate referral. The original LFT's were:

Bilirubin	58 μmol/l
AST	180 U/l
Alkaline phosphatase	140 U/l

What do you think?

81Q

A 25-year-old Asian male had a 5 year history of muscle cramps and paraesthesia of the extremities. He had lately been greatly troubled by painful spasms of the hands and feet. The physician to whom he was referred elicited Trousseau's and Chvostek's signs, and was not surprised when the following results came back:

Calcium	1.37 mmol/l
Phosphate	1.59 mmol/l
Albumin	48 g/l
Alkaline phosphatase	77 U/l
Urea	4.0 mmol/l

What is the likely cause and how would you confirm this?

80A

Hepatocellular jaundice

Hepatitis A is unusual in patients of this age, but nevertheless the LFT's are essentially hepatocellular—with moderate elevation of bilirubin, marked elevation of AST, and mild elevation of AP. Jaundice in men of this age is commonly obstructive, but the surgeon correctly referred the patient to a medical firm. He was admitted and had a variety of investigations including a liver biopsy. The eventual diagnosis (albeit one of exclusion!) was 'non-A, non-B hepatitis'.

81A

Hypoparathyroidism

Malabsorption and dietary deficiency of calcium or vitamin D must be excluded. The likely diagnosis in this case is hypoparathyroidism, with pseudo-hypoparathyroidism as a less likely alternative. The latter is due to end organ resistance to PTH, and can be differentiated (if necessary) by observing the response to infused parathormone—originally the 'Ellsworth–Howard test' (though nowadays levels of cyclic AMP rather than phosphate are measured in urine).

82Q

A 38-year-old man was admitted with abdominal pain. When gastroscoped, a duodenal ulcer was found and was assumed to be the cause of the pain. However, an amylase result subsequently returned very high, and acute pancreatitis was considered—though the pain had been present for some time. Subsequent amylase levels were as follows:

Day 1	1800 U/l
Day 3	1520 U/l
Day 7	1930 U/l
Day 10	1680 U/l

The man was free of pain from the 2nd day of admission. What is the explanation?

83Q

A 79-year-old woman was admitted deeply unconscious and extremely dehydrated. No history was available. The admitting doctor arranged some biochemical results, which returned as follows:

Na	K	Cl	CO_2	Urea	Creatinine	Glucose
155	5.1	120	23	16.5	163	43.0
mmol/l	mmol/l	mmol/l	mmol/l	mmol/l	μmol/l	mmol/l

What is the diagnosis, and what actions is needed?

82A

Macroamylassaemia

The permanently raised amylase levels suggest this rare but important abnormality in amylase structure. At further out-patient visits the serum amylase results were similarly high. The likely diagnosis of macroamylassaemia can be confirmed by differential amylase/creatinine clearance ratios, or absolutely by gel chromatography.

83A

Hyperosmolar non-ketotic diabetic coma

This is obviously a hyperglycaemic diabetic emergency, and the normal plasma CO_2 and anion gap (17) do not suggest ketosis. There is also hypernatraemia and hyperosmolality confirming 'HONK' coma as the problem. Plasma osmolality is best measured directly, but it can be fairly accurately predicted from the formula 2(Na + K) + urea + glucose, where all values are in mmol/l.

No urine was available from this patient when she presented. If it had, then heavy glycosuria without significant ketonuria would have been a further clue. A more sensitive (and immediately available) indicator of absence of ketosis is to test a sample of plasma for ketones by tablet ('Acetest') or strip '(Ketostix). This will be negative or only show a 'trace'of ketones present.

This patient was carefully rehydrated with half-normal saline and given a low dose insulin infusion. Despite this she died of a cerebral thrombosis.

84Q

An 18-year-old boy was investigated because of delayed puberty. The following hormone results were found:

Testosterone	5 nmol/l
FSH	1 U/l
LH	undetectable

What is the likely hormonal abnormality, and what further test is needed?

85Q

A 19-year-old insulin dependent diabetic was having frequent and severe hypoglycaemic episodes, despite gross reduction and eventually omission of insulin doses for a week. A co-existent insulinoma was considered and serum C-peptide levels were therefore taken during the hypoglycaemic episodes. The results, during 4 such episodes, were as follows:

glucose	C-peptide
	0.00 pmol/l
All less	0.00 pmol/l
than 1.0 mmol/l	0.01 pmol/l
	0.01 pmol/l
	0.00 pmol/l

What is the diagnosis?

84A

Hypogonadotrophic hypogonadism

The low testosterone, FSH and LH levels suggest a primary hypothalamic or pituitary problem. However a stimulatory test is needed to confirm failure of gonadotrophin release—the response of serum FSH and LH to injected LRH. This showed a flat response. Further pituitary tests are then indicated to establish whether there is hypofunction of other pituitary hormones (e.g. a TRH test, and the GH and cortisol response to insulin hypoglycaemia). In this case the FSH and LH undersecretion proved to be an isolated lesion. It was eventually mentioned by the patient that he had no sense of smell. This combination of congenital defects constitutes Kallman's syndrome.

85A

Factitious hypoglycaemia

In the pancreatic β cell, proinsulin splits to form insulin and C-peptide in equimolar amounts. In insulin dependent diabetics (where serum insulin levels are affected by insulin therapy) serum C-peptide concentrations may be a useful indicator of endogenous insulin secretion. The negligible C-peptide levels in this patient, at the time of hypoglycaemia, exclude insulinoma. The hypoglycaemia is due to injected insulin, and in the absence of it being given therapeutically—the diagnosis is factitious hypoglycaemia. This was eventually confirmed although the patient initially strenuously denied it.

Pituitary or adrenal failure may cause exquisite insulin sensitivity, but here the persistence of 'hypos' after a week's withdrawal of prescribed insulin makes self-administration most likely.

86Q

A 37-year-old lady was suspected of having acromegaly. GH levels were measured during a 50 g glucose tolerance test, with the following results:

time	glucose	GH
0 min	5.3 mmol/l	55.9 mU/l
30 min	7.5 mmol/l	62.9 mU/l
60 min	7.3 mmol/l	71.5 mu/l
90 min	7.7 mmol/l	65.5 mU/l
120 min	6.8 mmol/l	51.6 mU/l
150 min	6.3 mmol/l	48.0 mU/l
180 min	5.6 mmol/l	47.2 mU/l

Does this confirm the diagnosis, and what other investigative features are there of this condition?

87Q

A 40-year-old female was admitted with a 3 day history of severe abdominal pain. She had had 2 previous similar admissions, and nothing had been found to indicate a cause. Serum amylase levels had been normal or only borderline raised on both occasions, urinary porphyrins were normal and gastroscopy had been negative. On this occasion the amylase was as follows:

Amylase 462 U/l

The consultant read through the notes and decided that pancreatitis was still a possibility. Why did he think this, and what test did he perform?

86A

Acromegaly

The GH levels are markedly elevated and they fail to suppress during the GTT—this is the hallmark of acromegaly. Other biochemical features include impaired glucose tolerance, hyperphosphataemia, hypercalciuria and hypercalcaemia. Radiologically one may find an enlarged pituitary fossa, thickened heel skin pad and 'tufting' of the terminal phalanges.

87A

Late pancreatitis—raised urinary amylase

The elevation of serum amylase in pancreatitis may be brief, and especially if patients come in a little late in the illness, the levels may be normal or near normal. In these cases, the urinary amylase level may still be markedly raised—as was the case here. This test is made more accurate by estimating differential amylase/creatinine clearance. Serum lipase has also been suggested as a confirmatory test in this situation since activities decline more slowly than amylase after acute pancreatitis.

88Q

A fat and plethoric 31-year-old man was referred because of a variety of vague complaints including breathlessness, weakness and lethargy. On examination he was obese and had some active acne. The physician was unconvinced of Cushing's syndrome but decided it had best be excluded. He gave the patient 1 mg of dexamethasone to take at 12 MN and arranged for a plasma cortisol to be taken at 9 a.m. the next morning. The result was as follows:

Cortisol (9 a.m.) 400 nmol/l

What does this suggest, and what further tests are needed?

89Q

A 'brittle' insulin dependent diabetic had been vaguely unwell for some days, and began vomiting. He tried to drink and gave himself extra subcutaneous insulin when his blood glucose rose (according to home monitoring glucose test strips). The next day, however, he was admitted very ill and semi-conscious. Results were as follows:

Na	K	Cl	CO_2	Urea	Creatinine	Glucose
141	4.7	105	12	9.8	265	9.9
mmol/l	mmol/l	mmol/l	mmol/l	mmol/l	μmol/l	mmol/l

pH	7.32	Actual bicarbonate	11 mmol/l
pCO_2	2.8 kPa	Base excess	−13 mmol/l
pO_2	16.7 kPa		

What is the metabolic abnormality, how has it occurred, and how is it managed?

88A

Cushing's syndrome

This is the 'short dexamethsone suppression test', and the result shows failure to suppress plasma cortisol (assuming the dexamethasone was taken!) suggesting Cushing's syndrome. There is some argument over where the 'cut-off' point should be, but most normals will suppress to below 100 nmol/l. Some patients with endogenous depression suppress incompletely or not at all.

This test is useful to exclude significant autonomous secretion of cortisol, but elevated levels are not absolutely diagnostic. A full low and high dose dexamethasone suppression test as an in-patient is then indicated, with both cortisol and ACTH levels.

89A

Normoglycaemic ketoacidosis

This is classical diabetic ketoacidosis except that the blood glucose is normal or near normal. The electrolytes and blood gases are otherwise typical with low pH, low bicarbonate and base excess with raised anion gap. Note the elevated creatinine, disproportionate to the urea, due to assay interference from ketones.

Normoglycaemic ketoacidosis is a rare diabetic complication. It is poorly understood but most patients have tried to manage themselves at home too long, and have given insulin without carbohydrate and with insufficient fluid. Any persistent vomiting diabetic of course needs admission for intravenous therapy.

Once established, normoglycaemic ketosis is best treated with glucose/insulin/potassium infusion, as well as saline to correct dehydration.

90Q

An old man of 78 years was admitted as an emergency one evening. He lived alone and had no relatives, and was brought in by the police who had entered the house because he had not been seen for some time. There was no other history and on examination he was unconscious, had a normal temperature and was breathing deeply. Initial biochemistry was as follows:

Na	K	Cl	CO_2	Urea	Creatinine	Glucose
145	4.8	100	16	8.2	125	7.5
mmol/l	mmol/l	mmol/l	mmol/l	mmol/l	μmol/l	mmol/l

Urine glucose
Urine ketones   } negative by 'Labstix'
Urine protein

The admitting registrar performed arterial blood gases which confirmed a metabolic acidosis with hyperventilatory compensation. He sent a blood test off to the laboratory to test an idea he had had, and while waiting he did a side room test which provisionally confirmed the diagnosis. What was going on?

91Q

A 55-year-old lady was referred to an endocrinologist with '? thyrotoxicosis'. The physician thought her history and clinical signs were compatible with this diagnosis, but the biochemical results were rather confusing:

T4	120 nmol/l
FTI	110
TSH	< 1.0 mU/l (20 minutes after 200 μg of TRH, TSH was < 1.0 mU/l)

What might be the diagnosis?

90A

Salicylate poisoning

The problem is a severe metabolic acidosis, and ketoacidosis and uraemia have been excluded. Other possibilities are lactic acidosis and poisoning (e.g. by methanol, ethylene glycol or salicylate). Aspirin overdose is a not unlikely possibility and is easiest to test for before trying to get a lactate level. Whilst the lab was measuring the salicylate level, the RMO examined the urine. It has been negative for glucose on 'strip' testing (glucose oxidase method), but when he used the non-specific 'Clinitest' method for reducing substances the urine was heavily positive—salicylates are a well known cause of a false positive with this test. The lab soon returned a salicylate level of 780 mg/l, and a forced alkaline diuresis was instituted with good result.

91A

T_3 toxicosis

The T_4 is normal but the TRH response suggests hyperthyroidism. A serum T_3 was done and came back grossly elevated at 5.4 nmol/l. T_3 toxicosis makes up an uncertain minority of hyperthyroidism and should be remembered with thyroid function tests such as this. It should be remembered that the TRH test may be 'flat' without thyrotoxicosis in Cushing's syndrome (and steroid therapy), multinodular goitre, and ophthalmic Grave's disease.

92Q

The following results were obtained from a man of 35 admitted with persistent vomiting.

Na	K	Cl	CO_2	Urea	Creatinine
132	2.0	40	71	9.4	124
mmol/l	mmol/l	mmol/l	mmol/l	mmol/l	μmol/l

On the basis of these a diagnosis was made by the consultant clinical biochemist who saw the result, though it had already been made by the admitting medical registrar.

What was the diagnosis?

93Q

A 54-year-old lady was noted to have a plasma K value of 3.2 mmol/l 24 hours after a cholecystectomy. Her preoperative plasma K was 3.9 mmol/l. The following urinary electrolyte concentrations were subsequently found in a random urine specimen:

Urine Na	7 mmol/l
Urine K	106 mmol/l

Explain these values.

92A

Pyloric stenosis

The exceptionally high plasma CO_2 is likely to be due to metabolic alkalosis. The low Cl and K are due to loss of gastric secretions, compounded by the alkalosis. The urea is raised because of dehydration, and sodium and bicarbonate ions are conserved to maintain the extracellular volume. This compounds the metabolic alkalosis, as bicarbonate rather than chloride is absorbed in the proximal tubule as the counterion with sodium.

Such profound metabolic changes due to vomiting are generally only seen with pyloric stenosis. The most likely cause of this in a young man is scarring of a healed peptic ulcer, and this was later confirmed in this patient. The above description of the biochemical changes explains why the fundamental treatment is chloride replacement—usually in the form of liberal saline infusion.

93A

Secondary hyperaldosteronism

The reference ranges for urinary Na and K are wide and greatly influenced by diet. However, sodium excretion nearly always exceeds that of potassium. This pattern is reversed here, a change commonly seen in postoperative patients. This is due to secondary hyperaldosteronism, presumably caused by reduced blood pressure at operation activating renin secretion.

This investigation was inappropriate in this patient. The post-operative fall in plasma K was trivial and needed observation only. Also, if urinary electrolytes are to be measured, a 24-hour specimen is more representative.

94Q

A 36-year-old lady with a skin rash was admitted for investigation, when the following results were obtained:

Calcium	2.81 mmol/l
Phosphate	1.04 mmol/l
Albumin	39 g/l

After several days on cortisone 40 mg t.d.s., the plasma calcium was 2.23 mmol/l.

Explain the results.

95Q

A 25-year-old female was admitted complaining of extreme muscular weakness. Electrolytes and liver function tests were as follows:

Na	K	Cl	CO_2	Urea	Creatinine
144 mmol/l	1.7 mmol/l	93 mmol/l	42 mmol/l	3.3 mmol/l	53 μmol/l

Bilirubin	16 μmol/l
AST	123 U/l
Alkaline phosphatase	80 U/l

Explain these results.

94A

Sarcoidosis

Hypercalcaemia is most commonly caused by hyperparathyroidism or a malignancy. Less common causes include vitamin D intoxication, sarcoidosis and milk-alkali syndrome. The lowering of a raised calcium after administering cortisone is particularly characteristic of sarcoidosis, the diagnosis in this case, which was eventually proved by skin biopsy.

95A

Hypokalaemic myopathy

Severe potassium depletion may cause impairment of muscle function and ECG abnormalities (small or inverted T waves, U waves, Q–T interval prolongation). Gastric dilatation and paralytic ileus may result from changes in smooth muscle function. Weakness and, in extreme cases, paralysis may occur.

In this case the raised AST was from muscle rather than liver, since the serum CK was 3000 U/l and a myopathy was demonstrated by EMG. The plasma CO_2 was raised because of metabolic alkalosis secondary to hypokalaemia.

Hypokalaemia is usually due to increased gut or renal losses from various causes. This patient ate vast quantities of licorice, which contains glycerrhizinic acid—a substance with aldosterone like activity. She was treated with intravenous and oral potassium supplements, and advised to change her eating habits. At follow up her 'U & Es' were normal and she was well.

96Q

The following investigations were done on a 20-year-old girl with a 1-year history of vague abdominal pain. For the last 6 months she had noticed difficulty in moving her arms, and had pains in her fingers.

Urinary	aminolaevulinic acid (ALA)	289 µmol/24 h
	porphobilinogen (PBG)	39 µmol/24 h
	uroporphyrin	123 nmol/24 h
	coproporphyrin	67 nmol/24 h
Faecal	coproporphyrin	42 nmol/g dry weight
	protoporphyrin	189 nmol/g dry weight

What is the diagnosis?

97Q

A 45-year-old man was investigated for diarrhoea. Among the tests done was a urinary 5HIAA estimation.

Urinary 5HIAA   120 µmol/24 h

A sceptical consultant clinical biochemist rang the ward and established the cause of the result. Did the patient have the carcinoid syndrome?

96A

Acute intermittent porphyria

The symptoms are consistent with the neurological manifestations of porphyria. Elevated ALA and PBG levels suggest acute intermittent porphyria—this was confirmed by demonstrating reduced levels of erythrocyte porphobilinogen deaminase (uroporphyrinogen-I-synthetase). Uroporphyrin is formed non-enzymatically from PBG in urine and is responsible for the darkening which occurs with standing (port wine urine).

97A

Assay interference by paracetamol

The consultant was right to be cynical. The carcinoid syndrome is rare, but drug ingestion is common. Many drugs may interfere with 5HIAA assay, including paracetamol which this patient had received. When the test was repeated, it was normal. Paracetamol also interferes positively with glucose estimation using the Yellow Springs Analyser.

98Q

The following result was obtained on a child of 3 years with an abdominal mass:

Urinary HMMA   120 μmol/24 h

What is the likely diagnosis?

99Q

A 54-year-old lady was receiving L-thyroxine daily for hypothyroidism. For some time she had puzzled a number of registrars at the medical follow-up clinic, with her peculiar thyroid function tests. Clinically she was usually either euthyroid or possibly slightly hypothyroid. Typical results were as follows:

T_4 116 nmol/l
FTI   108
TSH   18 mU/l

What is the cause of these discrepant results?

98A

Neuroblastoma

Clinically, the diagnosis was thought to be either a Wilm's tumour or neuroblastoma. Urinary HMMA may be increased in the latter condition, and this diagnosis was confirmed histologically when a tumour was removed at laparotomy.

In an adult with a result such as this, phaeochromocytoma would be the provisional diagnosis. However, it must be remembered that colorimetric HMMA assays are subject to interference from drugs (and even some foods, such as bananas!).

99A

Irregular thyroxine therapy

The T_4 and FTI results are quite normal, yet the TSH level is elevated. Thyroid function tests which persistently show this abnormality in a patient on thyroxine replacement are strongly suggestive of poor compliance. In between visits to hospital patients may omit much of their therapy, but take tablets again shortly before their appointment. This can normalise the T_4 level, but not the TSH result, which takes longer to suppress.